BRIGHT WHITE LIGHT

Joseph C Jeluso

TABLE OF CONTENTS

ABOUT THE AUTHOR

I worked in a skilled trade most of my life, and had some college curriculum. I also enjoy writing and springtime is my favorite time of the year. The Bright White Light was somewhat, an idea from my studies with college electronics, and college curriculum brought this story to life. And I live with the most amazing lady I ever met.

ABOUT THE BOOK

Book number one is about a young boy who almost drowned at the age of ten, and his life story continues for another ten years.

At twenty years old, Joe and a co-worker witness the unbelievable sighting of an unidentified flying(floating) object and unexplainably saw the changes that happened.

The story concludes with a short explanation about the advances in technology and talks about how important it is to carry a camera.

CHAPTER 1

TOMATOES ON THE VINE

Once upon a time, in the kitchen of an adventurous and resourceful individual named Joe, a magical experiment unfolded. Joe, armed with curiosity and a desire to extend the life of his precious tomatoes, embarked on a culinary journey like no other.

In Joe's vibrant kitchen, where culinary experiments were constantly brewing, stood two resilient tomatoes. These brave fruits had managed to survive the passage of time, defying the natural course of decay. Unyielding in their quest to provide Joe with exceptional flavor, they perched on the countertop, their vibrant hues beckoning him.

However, as the days went by, the tomatoes started showing signs of fatigue. Their once plump and luscious bodies began to lose their firmness, a hint that their time was running out. Joe, unwilling to bid farewell to these last remaining warriors, resolved to take matters into his own hands.

With a stroke of inspiration, Joe recalled a fascinating technique he had come across—a secret whispered among seasoned gardeners and tomato aficionados. It was said that nurturing the vine even after the tomatoes had been plucked could bestow upon them an extended lifespan, keeping them succulent and bursting with flavor.

Taking a leap of faith, Joe grasped the vine firmly and allowed droplets of water to trickle down its tendrils, caressing the very source of life for the magnificent tomatoes. Day after day, he continued this ritual, tending to the vine with unwavering dedication. The tomatoes, bathed in this newfound sustenance, reveled in their unexpected reprieve.

Three days passed, and the transformation was astounding. The once weary tomatoes had undergone a remarkable rejuvenation. Their skin regained its tautness, their vibrant color returned, and their aroma intensified, permeating the air with a tantalizing fragrance.

Joe's heart leaped with joy as he beheld these revitalized treasures. It was a moment of triumph, a testament to the wonders of nature and the unyielding spirit of experimentation. Eager to savor the fruits of his labor, Joe embarked on a culinary escapade that would forever change his perception of the humble tomato.

Carefully slicing through the velvety flesh of one of the revived tomatoes, Joe marveled at its beauty. Placing it between two slices of artisan bread, he assembled a one-tomato sandwich fit for a king. As he took the first bite, a burst of flavor exploded on his palate—an orchestra of sweetness, tanginess, and earthiness, dancing harmoniously with each chew.

Overwhelmed by the sheer deliciousness of this simple creation, Joe couldn't help but exclaim, "Holy cow!" The tomato, nurtured by the life-giving waters of the vine, had transcended its humble origins, becoming a culinary masterpiece. From that day forward, Joe vowed to never let a tomato languish without the tender care of its vine.

And so, in the heart of Joe's kitchen, a tradition was born—a sacred ritual that breathed new life into every tomato that graced his countertop. Each watering session became a testament to his unwavering belief in the power of nature and the ability to transform the ordinary into the extraordinary.

The tale of Joe and his remarkable tomatoes serves as a reminder to all who embark on culinary adventures. Sometimes, the most delightful discoveries lie hidden within our everyday rituals, waiting to be unearthed and savored. With a touch of imagination and a willingness to experiment, even the simplest ingredients can be elevated to extraordinary heights, creating moments of pure gastronomic bliss.

CHAPTER 2

TIME, TECHNOLOGY, AND CHANGE

Born into the world in the wondrous year of 1951, I have been blessed to bear witness to an extraordinary era of human progress. The ceaseless march of time has woven together the tapestry of science and technology, unraveling the enigmatic secrets of the universe and transforming our very existence.

As I reflect upon the annals of history, my mind meanders to the awe-inspiring 1880s, where the tides of change began to swell. It was during this epoch that an ingenious mind, fueled by boundless imagination and a thirst for innovation, birthed a patent that would alter the course of transportation forever. A gasoline engine, harnessed with ingenuity and determination, beckoned a revolution in locomotion.

With the dawn of the early 1900s, humanity bid a fond farewell to the era of horse-drawn carriages, gracefully retiring them to the annals of nostalgia. The cacophony of hoofbeats gave way to the melodious purr of engines, as the masses embraced the newfound freedom bestowed upon them by the automobile. A convergence of gears, pistons, and wheels propelled us into an age of swift and efficient travel, forever transforming the landscapes we traversed.

Yet, the ceaseless pursuit of progress knows no bounds, and the march of technological prowess continues unabated. As if in a dance choreographed by the whims of destiny, we now find ourselves standing on the precipice of a new epoch—the advent of autonomous vehicles. The marvels of engineering and artificial intelligence intertwine, as our chariots of the future navigate the roads with a grace and precision only dreamed of by our forebears.

But such a breathtaking transformation is not limited to the realm of transportation alone. Cast your gaze upon the vast tapestry of human achievements, and you will witness the miraculous evolution of our species. From marveling at the flight of birds, we have soared through the firmament at unimaginable speeds, breaking through the sound barrier, hurtling forward at a breathtaking velocity of approximately 1,100 feet per second.

The realms of robotics, computers, global positioning systems, and the boundless expanse of the internet have woven themselves into the fabric of our lives, forging connections and transcending barriers.

With each passing moment, the curtain of knowledge is drawn back, revealing glimpses of the unfathomable wonders that lie beyond. It is in this realm of endless possibilities that the ethereal entity known as artificial intelligence beckons to us, its siren call tempting us to explore uncharted territories. What mysteries shall it unlock? What untold marvels shall it unfurl before our eyes? Only time can reveal the answers to these tantalizing questions, leaving us to surmise and dream of the limitless frontiers yet to be conquered.

As I bask in the golden glow of the years gone by, my heart brims with anticipation for the remarkable future that awaits. With every passing day, the symphony of science and technology crescendos, painting the world in hues of brilliance and wonder. It is our unwavering curiosity, our indomitable spirit of exploration, that propels us forward, daring us to reach for the stars and carve our names into the annals of history.

So let us embrace the evolving horizons of knowledge, for they hold the promise of a world yet unseen—a world where the boundaries of imagination dissolve, and the extraordinary becomes the norm. Together, let us embark on this intrepid journey into the realms of discovery, ever eager to unravel the enigmas that lay in wait, and to shape a future that transcends the limits of our wildest dreams.

CHAPTER 3

BETTER TO BE LUCKY

In the quaint embrace of a small town nestled in the heart of west Michigan, a tale unfolds, woven through the fabric of camaraderie and nature's allure. Within this close-knit community, I, the youngest of four brothers, found solace in the gentle rhythm of life. Born into this humble abode, I discovered my passion for angling, a timeless tradition passed down from my father, casting a net of fascination and adventure over my young soul. The convergence of two rivers on the southern fringes of our town held an irresistible allure, their crystalline waters whispering secrets that only a chosen few could decipher.

Catfish, bluegills, and bass danced beneath the surface, their sleek forms calling out to my eager heart. Their succulent flesh would grace our tables, rendering every meal a tantalizing feast for the senses. To this day, the memory of those delectable fish lingers on my palate, a testament to the enduring bond between a fisherman and his quarry.

In this realm of roots that ran deep, I discovered an enduring friendship in the form of Robert, a steadfast companion who shared my adventures and misadventures alike. Robert, affectionately known as Bob, and I were the epitome of yin and yang—complementary in our similarities and divergent in our differences. While my decision-making skills occasionally teetered precariously on the lower end of the good judgment scale, I possessed a spirit that would occasionally rise to the occasion, my luck often serving as a fortunate ally.

It was on one bitterly cold winter's day, when the air was crisp with the promise of frost, that I glimpsed the adventurous side of Bob. In our shared realm of imagination, he had bestowed upon me the moniker of Tarzan's boy, while he, in turn, bore the playful title of Bob. In the spirit of exploration and curiosity, we set out to traverse the icy expanse of the frozen river, located just a city block away from our homes. Across the snow-covered field we raced, our footsteps etching a temporary map upon the pristine canvas of nature's slumber. The skeletal trees, stripped bare of their emerald attire, stood sentinel in the midst of winter's embrace, while the dormant weeds whispered tales of seasons yet to come.

At last, we arrived at the river's edge, a point of convergence for the mighty Grand River, stretching approximately 200 feet wide and concealing unfathomable depths beneath its icy veneer. On this particular day, the river, often unforgiving in its temperament, lay encased in a formidable fortress of ice, over a foot thick. Awestruck by this frozen marvel, my eyes wandered upstream, where the ice had buckled and formed magnificent undulations, a testament to its formidable strength.

In the depths of my adventurous spirit, I turned to Bob and said, "Hey, Bob, let's venture to the other side." With an easy descent down a five-foot riverbank, softened by the erosion of time, we gingerly tested the ice beneath our feet, ensuring its strength. Our stomps resonated with an encouraging solidity, assuring us of a safe passage. And so, emboldened by the resilience of the frozen river, we traversed its icy surface, propelled by an insatiable curiosity that knew no bounds.

Once on the opposite shore, we found ourselves embraced by the silent grandeur of a barren woodland. A sense of exploration coursed through our veins as we embarked upon this uncharted territory, envisioning ourselves as intrepid explorers charting new frontiers. Alas, the winter landscape offered little in the way of hidden treasures, and our excitement soon gave way to playful mischief. Snowballs and small limbs became our main attractions, transforming the quiet woods into an arena of laughter and frolic.

But like all fleeting moments of joy, the time came for us to bid farewell to this snowy wilderness and make our return to the familiar shores. With unabated enthusiasm, we embarked on our journey back, running at full tilt. However, halfway across the icy expanse, fate revealed a treacherous twist. A small waterfall caught my eye, its frozen cascade beckoning me towards it. As Bob sprinted towards our point of entry, my adventurous spirit urged me to explore the icy marvel that lay before me.

In the face of this frozen tableau, strength coursed through the ice, but not uniformly. It yielded to the whims of a nearby creek, a mere six feet from the riverbank. And there, my forward momentum betrayed me. The ice beneath my feet gave way, plunging me into the icy grip of the river's relentless current. Panic seized my heart as I was pulled towards the fractured edges of the broken ice, its jagged shards poised to claim me as their unwitting prize.

But fate, ever a capricious companion, had other plans in store. In the midst of my desperate struggle, my gaze caught hold of a small branch, a lifeline emerging from the chaos. With a trembling hand, I reached out, my fingers closing around this slender savior. Against all odds, the branch held firm, resisting the ceaseless tug of the river's current. One hand at a time, I summoned every ounce of strength within me, pulling myself up the branch, inch by painstaking inch. And thus, against the relentless forces of nature, I emerged onto the muddy shore, gasping for breath and dripping with icy tendrils of water.

It was then that I noticed a small tree trunk, a beacon of hope within arm's reach. Grateful for this newfound ally, I gripped it tightly with my left arm, feeling the solidity of its embrace. Yet, despite my salvation, I found myself trapped upon the treacherous riverbank, its surface slick and unyielding. In this dire predicament, it was Bob, my steadfast companion, who extended a helping hand, his voice filled with concern and urgency. With a determined resolve, he slid me up the perilous incline, his strength proving instrumental in my safe ascent.

Soaked to the bone, with the chilling grip of winter clinging to my every pore, we ran tirelessly until the comforting warmth of our respective homes enveloped us. The magnitude of my escape settled upon me, the enormity of my luck cascading over my consciousness like a wave. Unbelievably fortunate to have narrowly escaped the clutches of the river's icy depths, I shuddered at the thought of what could have been—a dance with Hypothermia itself.

As I reflect upon that fateful day, I am reminded of the delicate balance between risk and fortune that shapes our lives. The smallest of decisions can lead us down paths unknown, and it is often the whims of chance that determine our fate. But in the midst of uncertainty, it is the enduring bonds of friendship that prove to be our steadfast anchor, pulling us from the jaws of danger and guiding us back to safety.

And so, the tale of Bob and Tarzan's boy, forever etched in the annals of our shared history, stands as a testament to the resilience of the human spirit and the enduring power of camaraderie. It serves as a reminder to embrace every moment with gratitude and to cherish the bonds that tie us to those who journey with us through the tapestry of life. In the end, it is the delicate interplay of fate, luck, and the unwavering support of kindred souls that shapes our destinies, weaving a tale that is uniquely our own.

CHAPTER 4

POWERFUL AND PERSUASIVE

Venturing into the rugged and ever-changing landscape of Michigan, my chosen vocation as an outdoor worker exposed me to a smorgasbord of weather phenomena that seemed to defy the boundaries of predictability. Raindrops cascaded from the heavens, morphing into delicate snowflakes as winter tightened its grip. The biting sting of sleet and the relentless embrace of sweaty heat served as constant companions throughout the seasons. Even in the face of brutal cold that seeped into my bones, I persevered, undeterred by the elements that sought to challenge my resolve.

As the early 1970s unfurled its wings, I found myself shouldering a new responsibility—the training of a crew of young men. In this role, I assumed the mantle of a problem solver, tasked with ensuring that each workday unfolded with efficiency, safety, and meticulous attention to detail. The weight of this duty rested upon my shoulders, as I strove to guide and inspire my team, nurturing a culture of excellence that would endure beyond the confines of a mere job. It was in these moments that I realized my chosen career was not just a means to an end; it was a lifelong pursuit, an achievement that would define the trajectory of my existence.

Yet, amidst the trials and triumphs of my professional life, I came to comprehend that hazards lurked not only within the realm of my chosen profession but also on life's myriad roads, both literal and metaphorical. The highways and byways that crisscrossed the landscape held their own hidden dangers, demanding my utmost vigilance. It was during my daily commutes, navigating the labyrinth of asphalt ribbons, that I experienced a couple of near misses, brushes with fate that served as stark reminders of the fragility of life. In those moments, the weight of mortality hung heavy in the air, cautioning me against the slightest lapse in judgment that could instantly transform a fleeting mistake into an irreversible tragedy.

However, as I cast my mind back to the early 1970s, an era when the roads were less crowded and the pace of life seemed somewhat gentler, I found solace in the relative rarity of such incidents. While a few flat tires and minor accidents punctuated my journey, they were mere blips on the grand tapestry of my existence. With an indomitable spirit and unwavering determination, I pressed forward, fueled by an unwavering belief that I was living the dream—a dream that seemed immune to the perils that lurked just beneath the surface. Oblivious to the hidden forces of nature that had taken root, I steered my course with an unwavering confidence, blissfully unaware of the challenges that awaited me on the horizon.

Yet, amidst the relentless pursuit of my professional aspirations, whispers of the extraordinary occasionally seeped into the realm of my consciousness, casting a spell of curiosity and wonder upon my mind. It was in the year 1968, when the airwaves of Detroit, the bustling metropolis affectionately known as the big D, crackled with tantalizing reports of flying saucers. Rumors swirled of a crash-landed extraterrestrial vessel in the desert southwest, while another was said to have been spotted hovering over the skies of Oklahoma. The radio waves trembled with excitement as broadcasters recounted tales of police pursuits, their patrol cars tracing the path of these mysterious objects, only to witness their astonishingly rapid evanescence.

These accounts, delivered with a mixture of awe and skepticism, ignited a spark of curiosity within me. I found myself captivated by the enigma that shrouded these otherworldly phenomena. Questions danced upon the periphery of my consciousness, begging to be explored. Could it be that we were not alone in this vast cosmos, that celestial visitors traversed the heavens, their presence eluding our comprehension? As the radio waves carried these tales of the extraordinary, I couldn't help but ponder the immeasurable possibilities that lay beyond the veil of our understanding.

In this symphony of life, where the mundane intertwined with the extraordinary, I continued my journey, propelled by the echoes of past experiences and the yearning for a future yet to unfold. The challenges that awaited me, both on the roads of Michigan and in the mysteries of the cosmos, would shape my character, test my resilience, and forever alter the course of my life's narrative. And so, with an open heart and an insatiable thirst for discovery, I embraced the winds of fate, ready to confront whatever hidden truths lay hidden along the roads less traveled.

CHAPTER 5

THE CAMERA

As the allure of the unknown beckoned, I sought to unravel the enigma of the hidden forces that shaped our world. Equipped with a camera that nestled within the confines of my trusty car, I embarked on a quest fueled by youthful bravery and an insatiable fascination. In the year 1970, I carefully stowed this photographic tool in the glove compartment, its film loaded and poised to capture the extraordinary—flying saucers and their elusive occupants. At the tender age of nineteen, I yearned for close encounters, eagerly anticipating the day when these round, metallic discs would unveil themselves before my lens. With every mile traveled, I held onto the hope that the skies above would reveal more unidentified flying objects, weaving tales of wonder and amazement.

Fate, in its whimsical dance, led me to a job location closer to my humble abode. A mere eighty miles east, the magnetic pull of those inexplicable forces drew me to an area in Michigan, traversing the winding corridors of highway M21. It was as if the very fabric of the universe conspired to guide me towards a nexus where my fascination with these ethereal crafts could find a tangible form. And so, my curiosity burned brighter than ever, igniting a fervent determination to capture the mysteries that danced across the firmament.

Armed with my camera, I set out on a personal mission. Each click of the shutter was a testament to my unwavering resolve, a declaration of my unwavering belief in the extraordinary. Should a saucer grace the earth with its landing, I would stand tall, shoulders back, chin held high, and document the momentous event through my lens. And if, by some stroke of cosmic fortune, the vessel's doors swung open, revealing its otherworldly occupants, I would seize the opportunity to immortalize their presence through the artistry of photography.

Yet, as the years unfolded, my meticulously crafted plan began to unravel, much like the delicate strands of time itself. The camera that once held the promise of capturing the unexplained lay dormant in my glove compartment, collecting a cloak of dust that bore witness to its neglected fate. The harsh

Michigan weather, with its icy grip in winter and scorching heat in summer, seemed to conspire against my ambitions, corroding the lens and freezing the film, rendering it useless. Doubt crept into my mind, whispering that my chance of encountering these flying metal apparitions had been forever lost.

Then, on a radiant summer day, a surge of determination coursed through my veins. As I meticulously cleaned the interior of my car, I made the fateful decision to part ways with my faithful camera. Its once-hopeful lens now marred by the passage of time, I bid farewell to the instrument of my dreams, resigning myself to the notion that the elusive encounter I had yearned for had slipped beyond the grasp of possibility. The radio stations, once abuzz with tantalizing reports of flying saucers, now fell silent, offering no new revelations or whispers of extraterrestrial visitors.

In the wake of this surrender, a profound sense of acceptance settled upon me. The allure of the unknown, while still nestled within the recesses of my soul, took on a quieter form. I found solace in the realization that the mysteries of the universe could not be contained within the limitations of a single camera or fleeting moments frozen on film. Instead, they resided in the boundless expanse of imagination, forever enticing us to explore the uncharted territories of what-ifs and maybes. And so, with the weight of unfulfilled aspirations lifted from my shoulders, I embraced the vastness of the world, knowing that even without tangible proof, the realm of possibility would forever ignite my spirit with its eternal flame of wonder.

CHAPTER 6
THE MOVE

In the realm of serendipity, amidst the toil and sweat of arduous workdays, fate weaved its intricate tapestry, intertwining the paths of two weary souls. It was a time when the amber hues of autumn painted the horizon, casting long shadows that mirrored the lengthening hours of our labor. Within the ebb and flow of conversations on those commutes, Jim, a fellow traveler in the realm of occupation, and I found solace in the idea of sharing an abode, a respite from the weariness of the road.

In the midst of September's embrace, when the days waned and dusk whispered of impending nights, we embarked on a quest to find refuge in the form of a rented sanctuary. The burden of long drives, coupled with the grueling demands of our work, propelled us to seek respite within the welcoming walls of an apartment. And so, we ventured into the heart of the town, embracing the opportunity to forge a new chapter in our intertwined lives.

Within the confines of that fully furnished abode, we crafted a routine that became the backdrop of our existence. Weekends and rainy days were the catalyst for our return to our respective homes, a brief intermission from the tapestry we were weaving together. But the weekdays, ah, they were a symphony of communal living. Breakfast and dinner were shared affairs, where we delighted in the flavors of the local restaurants, their culinary offerings providing nourishment for both body and soul. And as the sun retreated beyond the horizon, we bid adieu to the day's trials, finding solace in the embrace of sleep, awakening with the dawn's first light, ready to face another chapter of our shared odyssey.

Yet, unbeknownst to us, as we drifted in the currents of time, the unseen hands of destiny were subtly guiding our fates towards an unforeseen collision. The moments that lay concealed within the shroud of the unknown, fueled by a celestial countdown, were hurtling towards us with an inevitability that defied comprehension. The steady tick-tock of the clock reverberated in our ears, a constant reminder of the fragile nature of our existence.

But within the calm before the storm, we found solace in the routine we had crafted. Oblivious to the impending upheaval, we embraced the rhythm of our days, finding comfort in the familiar cadence of our shared existence. Little did we know that the foundations of our lives were trembling, paving the way for a convergence that would forever alter the course of our narratives.

As the hands of time continued their relentless march, we reveled in the simplicity of our routine, unaware of the approaching tempest. Our lives, entwined like ivy climbing a trellis, found solace in the familiar beats of everyday life. But destiny, with its capricious nature, had other plans in store for us. And soon, the ordinary would collide with the extraordinary, thrusting us into a world where time would no longer drift, but rather surge with an electric pulse, forever altering our perception of reality.

CHAPTER 7

WHAT IS IT?

As the gentle touch of autumn embraced the land, casting a golden glow upon the landscape, our weary bodies yearned for respite after a long day's work. Fatigue and hunger intertwined, urging us to seek nourishment and rest within the haven of our apartment. The day had been like any other, following the familiar rhythm of labor and sustenance.

A quaint little restaurant nestled on the south side of M21 beckoned us with its warm glow, promising a refuge from the outside world. Its charm lay just a couple of blocks west of the bustling downtown, where towering buildings etched their silhouettes against the dusky sky. Parking our vehicle with ease, we made our way towards the entrance, our hunger piqued and anticipation growing.

Inside, the aroma of home-cooked meals enveloped us, embracing our senses with a comforting familiarity. Taking our seats, we ordered our usual fare, the dishes that had become synonymous with our visits to this establishment. Conversation flowed effortlessly as we savored each bite, the weariness of the day gradually fading away.

With satisfied appetites and thoughts of winding down, we paid our respective bills, the mundane transaction marking the end of another day's journey. Exiting the restaurant, we ventured back to my car, its metal shell shielding us from the outside world. I took my place behind the wheel, rolling down the window and igniting a cigarette, a moment of respite before embarking on the final leg of our day.

As we cruised along the familiar stretch of M21, our eyes were drawn to something peculiar. Two separate gatherings of people stood upon the sidewalks, their gazes fixed upon the western horizon. Curiosity piqued, we exchanged bewildered glances, wondering what had captured their attention. Leaning out of the window, I strained to catch a glimpse of the enigma that eluded our sight. The towering buildings obstructed our view, leaving us in a state of perplexity.

A shared sense of intrigue propelled us to seek answers, and with a mutual agreement, I executed a swift right turn, embarking on a hasty U-turn maneuver. The clusters of onlookers remained rooted in their places, their gazes locked in the direction we had just come from. We rejoined the flow of traffic, curiosity now driving our actions.

Continuing our westward journey along M21, it didn't take long for the mysterious spectacle to reveal itself. With a mixture of awe and bewilderment, we beheld a small, luminous white light in the distance. It hovered in the vast expanse of the sky, defying our understanding of the world around us. Exchange of glances and whispered questions hung in the air, but neither of us could fathom the nature of this ethereal presence.

Driven by an insatiable desire to unravel the mystery, we made a unanimous decision to embark on a pursuit. The daylight offered us ample visibility as we traversed the county road, our eyes fixated on the enigmatic light that defied explanation. The road stretched before us, devoid of other travelers, as if the universe had conspired to grant us sole witness to this otherworldly phenomenon.

Miles slipped away beneath our wheels as we chased the mysterious light, its steady presence never faltering. The remote landscape unfolded before us, a testament to the vastness of the unknown. Without hesitation, we pressed on, our curiosity fueling our determination.

Finally, beneath the radiance of the light, we reached an intersection. Acting swiftly, I maneuvered the vehicle onto a gravel road, finding solace in its wide shoulder. Coming to a stop, we emerged from the car, our gazes fixed upon the celestial enigma suspended above us. Its brilliance mirrored that of a fluorescent light, resembling a small moon while defying all conventional explanations. It hung motionless, a circular entity that defied gravity and comprehension.

Time stood still as we stood there, observers to an unexplainable phenomenon. The light emitted no discernible movement, evaded any comparison to a mere spotlight, and remained devoid of a discernible glow. Estimating its distance to be around a mile and a half, we found ourselves captivated, unable to tear our eyes away from this enigmatic presence. As we stood by the fence, the moments stretched into minutes, the weight of the unknown hanging in the air.

The descent of the sun began, casting its farewell glow upon the horizon, and the light intensified. We exchanged glances, acutely aware of the approaching dusk. Peripheral vision constantly aware, we pondered our next course of action.

"Should we depart?" I questioned, seeking Jim's opinion. His response was a nonchalant dismissal, "Nah, it's not doing anything." With shared uncertainty, we took our first steps away, our eyes lingering on the ethereal sight above.

Suddenly, as if triggered by the fading light, the unthinkable transpired. The red light flared to life, propelling the white vessel towards us, a breathtaking display of movement and power. In an instant, it felt as if we had plunged into icy waters, the shock of the unexpected sending us into a frenzied escape. Vaulting over the fence, we ran for our lives, seeking refuge behind the sheltering embrace of two trees nestled within the nearby woods. There, hidden from immediate danger, we observed the unfolding spectacle with bated breath.

Beneath the bigger white light, the red one hovered, suspended in an eerie stillness. The world seemed to hold its breath, anticipation hanging heavy in the air. Without warning, the red light ascended, joining another red light a couple of blocks above the ground. Two crimson orbs defied gravity, their presence defying logic.

Mesmerized and bewildered, we watched as the white vessel erupted in a cataclysmic explosion. Yet, no sound reached our ears, and we observed no debris hurtling through the air. Moments later, the red lights followed suit, ascending into the sky, their crimson glow transforming into pure, blinding whiteness. And just like that, they vanished, leaving nothing but a fading memory etched into the fabric of our minds.

Emerging from the concealment of the woods, our hearts heavy with disbelief, we found our way back to the car. What had we just witnessed? The weight of the experience settled upon us, leaving us grappling with a reality that seemed to defy explanation. Nightfall descended, shrouding the world in darkness as we retraced our path along the county road, heading towards M21.

As the rhythmic hum of the tires accompanied us, the extraordinary refused to relent. A meteor, ablaze with a kaleidoscope of colors, streaked across the sky to our right. Its vibrant display painted the heavens, leaving a fading trail of hues in its wake. The ephemeral beauty vanished within seconds, leaving us in awe of the extraordinary night we had just borne witness to.

Silence enveloped the car, two minds attempting to comprehend the incomprehensible. The night held its secrets, and we could only drive on, haunted by the unforgettable encounters that had unfolded before our eyes. The events of that evening had forever altered our perception of the world, leaving us with a profound sense of wonder and a longing for answers that would elude us in the years to come.

CHAPTER 8

HEY BRO

Throughout the years, I found myself unable to shake the memory of that extraordinary encounter. Each passing day, spanning half a century, bore the weight of contemplation as I delved deeper into the enigma that had etched itself into my consciousness. And amidst the countless moments that wove the tapestry of my life, certain revelations began to surface, like faint whispers in the wind.

One question, posed by my oldest brother in the late 1970s, echoed in my mind throughout the years. "How did the bottom open?" he had asked, his curiosity mirroring my own. At the time, I had no answer to offer him, my thoughts clouded by uncertainty. All I could muster was a feeble response, suggesting that perhaps two sections slid outward. But the truth eluded me, slipping through the crevices of my understanding.

It wasn't until the early 1990s that a glimmer of insight began to dawn upon me. Through logical deduction and countless hours of contemplation, I pieced together a plausible explanation. The bottom had opened with such astonishing speed that even my perceptive senses failed to capture its movement. It must have unfolded like the crack of a whip, a swift and precise motion executed in a mere second. Such advanced technology, beyond our comprehension, must have been at play, utilizing high-speed motors and intricate mechanisms to orchestrate the opening of the hatch.

Driven by an insatiable hunger for knowledge, I immersed myself in the vast reservoir of scientific shows and documentaries. I embarked on a journey of self-education, exploring the realms of electronics during my college years. The skies became my canvas, my watchful gaze fixated upon the celestial expanse that had once unveiled a sight beyond imagination.

Years turned into decades, but my quest for understanding never wavered. I became a student of the cosmos, absorbing every ounce of information that could shed light on the mysteries that had unfolded before me. From astrophysics to aerospace engineering, my pursuit of knowledge knew no bounds. And with each passing day, the fragments of the puzzle began to align, gradually forming a tapestry of comprehension.

Yet, even with my newfound insights, questions still lingered, like echoes reverberating through the corridors of my mind. What lay behind the technology that propelled the floating light? What civilization or force had harnessed such boundless power and mastery over the laws of physics? The answers remained elusive, lost in the vast expanse of the unknown.

Half a century had slipped through my fingers, but the memory of that fateful night burned bright within me. It served as a constant reminder of the unfathomable depths of the universe, a testament to the infinitesimal sliver of knowledge that we, mere mortals, possess. And though the journey had yielded some semblance of understanding, the true essence of that extraordinary encounter continued to elude me, leaving me forever captivated by the grandeur of the cosmos and the infinite possibilities that lie just beyond our reach.

CHAPTER 9

SUN SWITCH

Within the confines of the white flying saucer, an intricate operating system hummed with the power of electricity. It was a marvel of advanced engineering, designed to navigate the skies with seamless precision. But like any technological marvel, it required a source of energy to sustain its operations. And what better place to harness this energy than the Earth's own magnetic field?

As the white saucer traversed the heavens, it tapped into the invisible currents of the planet's magnetic embrace, drawing power to replenish its reserves. The magnetic field acted as a conduit, allowing the saucer to recharge its systems, ensuring that it could continue its celestial journey without interruption. However, as the sun dipped below the horizon, the power supply waned, and a crucial switch was triggered.

In those fleeting moments when daylight surrendered to darkness, a transformation unfolded within the white saucer. The weakening power supply necessitated a strategic maneuver—a calculated adjustment to conserve energy and maintain the delicate equilibrium of its systems. And so, with a precise mechanism set in motion, the bottom of the saucer opened.

Within the vessel's compact interior, space was a premium commodity. It had to accommodate not only the two red flying saucers but also a myriad of intricate components and enigmatic technologies that defied comprehension. The limited quarters within the white saucer hinted at the advanced miniaturization and efficient use of space employed by its creators.

Reflecting upon these revelations, I found myself indebted to my brother Vince. His probing questions had inadvertently unearthed a valuable piece of the puzzle. While the specifics of the white saucer's inner workings remained shrouded in mystery, the notion of utilizing Earth's magnetic field for recharging and the intricate dance of opening its bottom began to align.

Yet, as one question found its answer, a cascade of new inquiries flooded my mind. What purpose did the red saucers serve within the white vessel? What other remarkable technologies lay concealed within its compact confines? The enigma continued to deepen, drawing me further into its captivating embrace.

In my quest for understanding, I embarked on a relentless pursuit of knowledge. I delved into the realms of electromagnetism and power systems, seeking to unravel the intricacies of the white saucer's operations. The interplay between magnetic fields, electrical currents, and intricate circuitry captivated my attention, offering glimpses into the inner workings of this extraterrestrial marvel.

As the years rolled on, my fascination with the enigmatic flying saucers remained unabated. It became clear that these awe-inspiring vessels were far more than mere spacecraft—they represented the pinnacle of technological achievement and a testament to the vast depths of the universe's wonders.

So, armed with newfound insights and fueled by unquenchable curiosity, I delved deeper into the mysteries that lay before me. The white saucer, its red companions, and the intricate web of systems that governed their operations beckoned me to unravel their secrets. With each revelation, I inched closer to comprehending the unfathomable realities that existed beyond the boundaries of our world.

The journey was far from over, and as I pursued the truths hidden within the confines of that extraordinary encounter, I remained forever grateful to my brother Vince for inadvertently shedding light on one aspect of the enigma. The tantalizing puzzle continued to expand, drawing me into its gravitational pull, leaving me eager to uncover the next chapter in this extraordinary tale.

CHAPTER 10

SPEED

Over the course of several decades, a persistent question gnawed at the corners of my mind, urging me to seek answers. It pertained to the astonishingly slow speed at which the flying saucers traveled. Considering the vast distances between star systems, it seemed virtually impossible for these vehicles to embark on interstellar journeys. The sheer enormity of time it would take them to reach even the nearest star left me in awe.

As the years passed and technology advanced, the advent of social media brought about new avenues for exploration. Through the digital realm, I stumbled upon a captivating film captured from an airplane—an unsuspecting observer had recorded a flying saucer in close proximity to their aircraft. To their astonishment, the saucer vanished into thin air, mirroring the behavior of the two saucers I had encountered all those years ago. One turned white and disappeared, followed by the second, leaving me to believe they had simply moved beyond my visual range.

Yet, the laws of science suggested something intriguing—a thought that had previously seemed inconceivable began to take root in my mind. It became apparent that these remarkably advanced vehicles possessed a capability far beyond conventional propulsion systems. Could it be that they had at their disposal a technology capable of triggering a stargate?

The concept of a stargate had long been relegated to the realms of science fiction, dismissed as an impossibility. But as I immersed myself in the realm of possibilities, I started to entertain the notion that these extraordinary saucers could indeed manipulate space-time to traverse vast cosmic distances. The very fabric of reality seemed to be at their command.

In this newfound line of thinking, the disappearance of the saucers took on a whole new significance. It became evident that their departure was not a result of distance alone, but rather a deliberate activation of an otherworldly mechanism. The saucers, guided by their advanced technology, could seamlessly transition between dimensions or create portals that bridged the gap between celestial realms.

This revelation propelled my mind into a realm of speculation, fueling a myriad of questions. What lay beyond those gates? What realms awaited exploration on the other side? And what kind of extraordinary metal could be harnessed to illuminate their paths, defying the laws of physics as we understood them?

It was a tantalizing journey of speculation, guided by the flickering glow of newfound understanding. The enigmatic saucers that had once puzzled and fascinated me were now imbued with a sense of purpose—a purpose that extended beyond mere interstellar travel. They were the gatekeepers of possibilities, bearers of knowledge and exploration that transcended the limits of human comprehension.

With every passing moment, my conviction grew stronger, and the limitations of conventional understanding crumbled before the enigmatic capabilities of these advanced beings. The world as I knew it expanded, and the boundaries of what I once believed to be possible dissolved, replaced by the boundless wonders that lay beyond the veil of the unknown.

As I ventured deeper into this labyrinth of speculation, I knew that I was standing on the precipice of discovery. The secrets of the saucers, their method of travel, and the hidden realms they traversed awaited my unwavering pursuit. With each piece of the puzzle I uncovered, I drew closer to unraveling the true nature of their existence—a journey of awe, wonder, and infinite possibility that would forever alter the course of my understanding.

CHAPTER 11

WHAT'S NEXT?

The future, a vast tapestry of untold possibilities, beckons us with its enigmatic allure. Within its realm, lies the key to unraveling the mysteries that elude us today. As the wheels of time continue their ceaseless march forward, they hold within them the answers we seek, gradually unveiling the secrets of technological advancement.

In the vast expanse of the unknown, my two questions linger, suspended like whispers carried by the winds of uncertainty. Their answers, elusive and intangible, dance just beyond the reach of my senses. The origin of these enigmatic flying saucers and the destinations that beckoned them remain shrouded in a veil of enigma. However, time, that eternal revealer of truth, patiently bides its time, waiting to bestow upon us the wisdom we yearn for.

Consider the remarkable journey humanity has undertaken in the span of a mere 140 years. From the humble back of a horse, we have ascended to the skies, soaring through the heavens in majestic jets. The strides we have made in science, technology, and innovation have reshaped our world in unimaginable ways, stretching the boundaries of human achievement.

Yet, as we stand on the precipice of the next century, we are left to ponder what marvels lie just beyond the horizon. The unfolding chapters of human progress hold within them the promise of even greater wonders. With each passing year, the inexorable march of innovation propels us towards realms yet unexplored, offering tantalizing glimpses of what lies ahead.

But amidst the vast panorama of future possibilities, one thing remains certain—our phones, those compact devices of communication and connection, will undoubtedly continue to play an integral role in shaping the narrative of our lives. Armed with their powerful cameras, we stand ready to capture fleeting moments of awe and fascination should the light of the saucers grace our skies once more. Our phones, the silent witnesses to the extraordinary, poised to document the extraordinary with a single click.

As time unfurls its tapestry, we embark upon a journey of discovery, guided by the boundless curiosity that defines our human spirit. The next century promises to be an era of breathtaking innovation, where the limits of human potential will be stretched to their very boundaries. In the relentless pursuit of progress, we will unravel the secrets of the universe, unearthing knowledge that will reshape our understanding of the cosmos and our place within it.

So, let us embrace the unfolding future with open hearts and inquisitive minds. For within its embrace lie the answers we seek, the possibilities we dare to dream, and the extraordinary tales that will be etched into the annals of history. As time unravels its enigmatic riddles, we will bear witness to the remarkable journey of humanity, forever driven by our insatiable thirst for knowledge and our unyielding determination to push the boundaries of what is known.

And so, we await the revelations that lie ahead, ready to traverse the uncharted realms of tomorrow, armed with nothing but our unyielding curiosity and the timeless allure of the unknown. The future beckons, and we, as seekers of truth, stand poised on the precipice of discovery, eagerly anticipating the wonders that await us in the yet-unwritten chapters of our shared human story.

CHAPTER 12

TOMATOES OFF THE VINE

Tomatoes, those vibrant orbs of nature's bounty, possess a transient beauty that captivates both the eye and the palate. However, their ephemeral nature demands swift action to preserve their succulence, unless we turn to the art of canning. Beyond their mere aesthetic allure, tomatoes harbor within them a treasure trove of essential nutrients, offering a wholesome blend of carbohydrates and proteins that nourish our bodies.

Nestled within the heart of each tomato, hidden within its delicate seeds, lies a veritable cornucopia of goodness. These tiny capsules of life boast an array of vital nutrients, delivering a satisfying dose of energy, along with the fortification of carbohydrates and proteins that sustain us in our daily endeavors. Moreover, tomatoes grace us with their modest calorie content, making them a guilt-free indulgence, while their low-fat composition aligns with our pursuit of a balanced and healthy lifestyle. As if that weren't enough, tomatoes generously bestow upon us the gifts of vitamin C and vitamin K, augmenting our immune system and supporting healthy blood clotting, respectively.

In the delightful company of my dear companion, Lizz, we revel in the joys bestowed by nature's bounty, particularly in the form of tomatoes clinging tenaciously to their vines. A delightful dance unfolds as we cultivate these vines with tender care, nurturing them with water and devotion, knowing full well that their rewards will manifest in the long run. And when the time is ripe, quite literally, we partake in the wondrous pleasure of savoring homegrown tomatoes, basking in their unparalleled flavor and inherent benefits.

Ah, the first vine, a testament to our efforts and patience. As the days passed, its fruits ripened with anticipation, and when we finally plucked that first tomato from its nurturing embrace, we were greeted with a burst of delightful flavors that titillated our taste buds and gratified our palates. The experience was nothing short of sublime, a symphony of taste and texture that left an indelible mark on our culinary memories.

Yet, as time marched on, the subsequent tomatoes on the vine beckoned, their alluring appearance enticing us with promises of gustatory delight. Alas, though they retained their firmness and appeared resplendent in their external beauty, a disheartening truth unveiled itself. Time had slowly eroded their once-vibrant flavors, leaving behind a mere shadow of their former glory. We had unknowingly allowed those precious tomatoes to languish a day or two too long on the kitchen countertop, their essence fading away with each passing moment.

Reflecting upon this experience, we gleaned a valuable lesson in the art of tomato stewardship. We realized the importance of promptly savoring these delectable fruits, for their ephemeral nature demands swift action. Each tomato, a fleeting masterpiece of nature's craftsmanship, is best enjoyed at the pinnacle of its flavor and ripeness. It is a delicate balance, a harmonious interplay of timing and appreciation, as we strive to capture the essence of the tomato's fleeting brilliance before it dissipates into the annals of time.

And so, armed with this newfound wisdom, we embark on our culinary adventures, cherishing the ephemeral delights that nature gifts us. In the realm of tomatoes, we partake not only in their tantalizing flavors but also in their nourishing embrace. For within their luscious flesh lie the nutrients that sustain us, the vitamins that invigorate us, and the undeniable joy that accompanies the act of savoring nature's bountiful offerings.

Made in the USA
Coppell, TX
05 November 2026

11302307R00050